Abstract

The aim of this project to find out the different factors affecting buying intensions of customers towards smart phones . A conceptual model is developed that extrinsically and intrinsically affect the purchase decisions of the customers. An extensive use of literature review is considered for better analytical research. The research aims to overcome the gap of understanding the concept of buying intensions for Smartphones which motivates customers in making the purchase decision. This study provides valuable insight into consumer behavior regarding Smartphones demand by examining the factors that influence customers demand for using and owning them. The findings of the study identifies Product Feature, Price, Social influence, Brand Name and Convenience factors which affects the buying motives for Smartphone.

Introduction

Smartphones have made existences of individuals much easier and comfortable. Innovation is the essential explanation behind getting comfort into individuals everyday life. It has enhanced the expectations for everyday eases and individuals' way to deal with things. India is challenging the worldwide pattern in Smartphone Market since it is still a developing business segment an expanding number of individuals are upgrading from feature phone to a smartphone. More than 70 percent of cell phone smartphone users were relied upon to move up to premium smartphone models. An expanding number of smartphone companies took an interest in the Make in India activity, and domestic manufacturing boosted sales of smartphones in India.

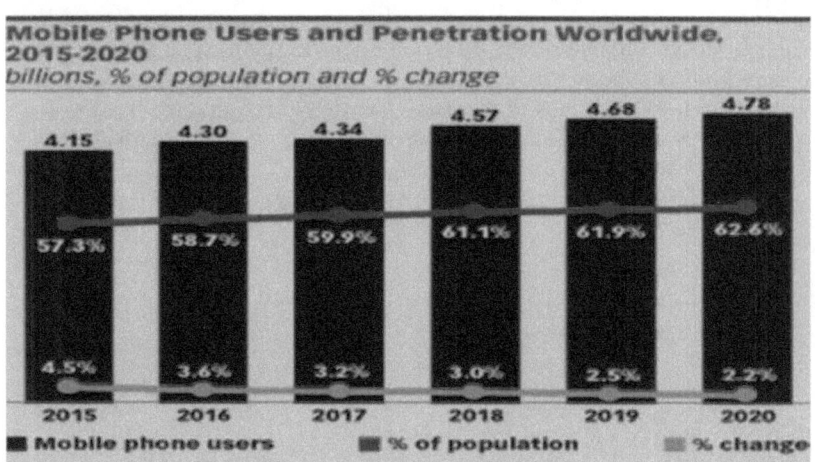

Source: As explored in a new eMarketer report, published on November 23, 2016

Whereas the move from featured phone to smartphones is stirring in many markets around
the world, the rate at which it is going on varies from district to area. Generally half (49.7%)of cell phone clients over the globe will utilize a cell phone at any rate once per month this year, It appraises that that there will be 4.30 billion Smartphone customers worldwide in 2016, discourse to 58.7% of the worldwide population. Despite the fact that shopper

development is moderating, the quantity of Smartphone customer will move to 4.78 billion in 2020 . India secured the second spot in the list of the world's largest smartphone market, enlisting in at a healthy growth rate of 23% Year over year. As per report from Counterpoint(April 2016) , there is an opportunity for growth in Indian smartphone In India more than 1.2 billion individuals have smartphone in the hands of less than one fourth of a billion of its people.

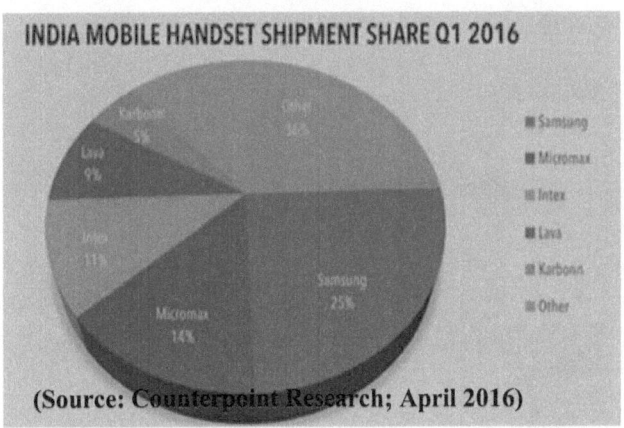

(Source: Counterpoint Research; April 2016)

Research Objectives

- ❖ To explore the previous studies related to purchase intentions of consumers towards smartphone.

- ❖ To identify the key factors that motivates consumers to purchase and use Smart Phones.

- ❖ To explore factors affecting perceived obsolesce with respect to Smartphone.

Literature Review

On the basis of investigation, first of all study describes the trends in the mobile sector in order to illuminate the issues underlying consumer behavior. This will be followed by review of recent studies concerning factors that seems to affect the choice of a mobile Phone, operator as well as intentions to adopt new mobile phone features and services such as multimedia messaging services and sending emails. The author conducted an extensive review on literature on national and international perspectives of mobile phone consumer purchase intention/behavior of last 15 years. The finding of the literature reviews were based on search of keywords like "purchase intentions", "influencing factors", "consumer purchase decisions", "product features". Finally paper reports the survey findings and ponders managerial and theoretical implications. The mobile phone is not only a device of communication but it is an in trend gadget which depicts individual's status, identity, style, and pattern (Deng, 2006) . Currently Mobile phones are named as 'smartphone', as they offer more advanced computing power, advanced features and connectivity. Furthermore the smartphone's major capabilities to make voice call, video call, SMS, and MMS, smartphones have been transposed as a "new information medium" (May, H., & Hearn, G. 2005).

A brief history of Smartphones

It seems as though just about everyone owns a smartphone, even though it wasn't that long ago that the technology was first introduced to the general public. With their advanced computing capabilities and other features, smartphones have quickly gained popularity. Prior to the invention of smartphones, there were several devices that were used including regular mobile phones, and PDA devices. Eventually technology was combined and the concept of the smartphone was born. The first concepts that eventually led to the invention of the smartphone date back to the 1970s. It was not however until 1992 that IBM came up with a prototype mobile phone that incorporated PDA features. The device was demonstrated the same year at a computer industry trade show called COMDEX. In 1994, BellSouth came up with a more refined version of the phone and called it the Simon Personal Communicator. This is the first device that could really be referred to as a smartphone. The Simon Personal Communicator was able to make and receive calls, send faxes and emails, and more. In the later part of the 1990s, many mobile phone users began carrying PDAs (personal digital assistants).

Early PDAs ran on various systems such as Blackberry OS, and Palm OS. Nokia released a phone combined with a PDA in 1996. The device was called the Nokia 9000. In 1999, a smartphone was released by Qualcomm. The pdQ smartphone, as it was called, featured a Palm PDA with internet connectivity capabilities. Smartphone technology continued to advance throughout the early 2000s. These advances in technology brought about the introduction of the iPhone, the Android operating system and more. From 2011 to current times, smartphone technology has become even more advanced and technology continues to improve at a rapid pace. Smartphones have more features than ever before including touchscreen capabilities, GPS, cameras, and much more.

Market Share

According to Faisal Kawoosa, Lead Analyst, Telecoms at CMR, "As we see Smartphones becoming the necessity in India, the significance of high-end Smartphones only amplifies for the reason that these many users would like to go for an upgrade from mid-level Smartphone to a higher order Smartphone."

Post-2013, we saw Smartphones booming in India and a fair installed base is now ready for replacement, and our observation is that approximately 70% of the users would go for an upgrade to a higher end Smartphone rather than the replacement in the same price segment. The upgrade is in terms of functionality, feature set and of course the price.

The key players in the high-end smartphone brands are:

Samsung: Samsung retained its number one position in the worldwide smartphone market with a 21% share in 2016Q3. This retention of position comes in the midst of the Galaxy Note 7 recall, thanks to the continued success of the Galaxy S7 and S7 Edge devices. Samsung's streamlined portfolio of devices, including the affordable J-series, proved successful in many mid-tier markets that were typically dominated by local brands. How fast Samsung recovers from the damage to its brand, remains to be seen.

Apple: Apple shipped a total of 45.5 million units, which is a 5.3% decline year on year from 2015. The new iPhone SE did well in both emerging and developed markets. The iPhone 6S continued to be its bestselling device this quarter, followed by its newest device, the iPhone 7 and about to launch iPhone8.

Sony: Vijay Singh Jaiswal, head of Xperia business of Sony India said that Sony India's marketing strategy will focus on offering premium products in the smartphone segment. Sony Mobiles just has around 1.5 percent market share in India. According to IDC, Sony's share fell from 3.5 percent in 2014 to 1.4 percent in 2015. This year, in India, around 24.4 million smartphone units were shipped in 2016, which is a 12 percent year on year growth.

HTC: The handset maker Thursday launched seven new 4G smartphones, including the flagship HTC 10 and a new Desire handset line-up, which already makes up 80% of the overall sales in India. The company now has 15 smartphones in its portfolio, starting from Rs 8,000 to Rs 52,990, 12 of which support 4G technology. The company, as per its internal findings, claims around 4% volume share in the smartphone market, which it wants to increase to 10-12% in near future.

OnePlus: When the OnePlus One was launched in India in 2014, only 15% of the smartphones were sold online. That share jumped to 37.3% in 2015, according to research firm IDC. Even though the flagship smartphone space has clear winners in many countries, India still remains up for grabs. Pei points out that OnePlus has already captured 7% of the market share in India

An analysis on the Top 5 Smartphone's future

Samsung

Status: The undisputed market leader of smartphone market.

Strengths : It has remarkably been able to manage its position across different segments of mobile handset market. A very strong offline footprint.

Challenges: With too many challenges at different levels , will be difficult for it to retain market shares even continuing to be the leader.

Outlook: Positive – will manage to be the leader in the market, thanks to its wide portfolio, established reach and brand recognition.

Apple

Status: Strong growth in 2015 with no direct competiton being the sole player in its ecosystem.

Strengths : very high Aspiration Value and premium tag.

Challenges: Will not be able to gain in 20-30k range, which will have the maximum growth in high-end Smartphone segment.

Outlook: Positive- With its brand value and strong pull is expected to grow 50% in 2017.

Sony

Status: Diminishing market status year on year.

Strenghts: Long time established brand operating at attractive price points.

Challenges: Overall declining market status in the smartphones. Have been slow on new products and ranges

Outlook: Negative-With no revolutionary strategy around. Sony will have a major set back in this category as well. Need to revisit the entire value chain for a turnaround.

HTC

Status: A niche player and within top 5 in this segment.

Strenghts: Premium image of the brand and global acknowledgement for designs.

Challenges: As it dilutes further towards the medium range Smartphones managing the premium range would be difficult.

Outlook: Stable- With its premium range, will continue to gather market standing as not many players will look at this category of smartphones, limiting the competiton.

OnePlus

Status: Already among top 5 players in 2016 with growing market share.

Strenghts: Loyal tech community, attractive pricing, growing online channel share, differentiated positioning in premium segment, innovative marketing strategies.

Challenges: Limited Distribution reach due to no offline presence, low brand recall outside towns and cities.

Outlook: Positive- With Its OnePlus 3t on the the market. OnePlus is likely to grow its category share further in 2017

Elements that prioritize purchase decision:-

- ➢ Price
- ➢ Product Features
- ➢ Social Influence
- ➢ Dependency
- ➢ Compatibility
- ➢ Convenience
- ➢ Brand Name
- ➢ Relative Advantage

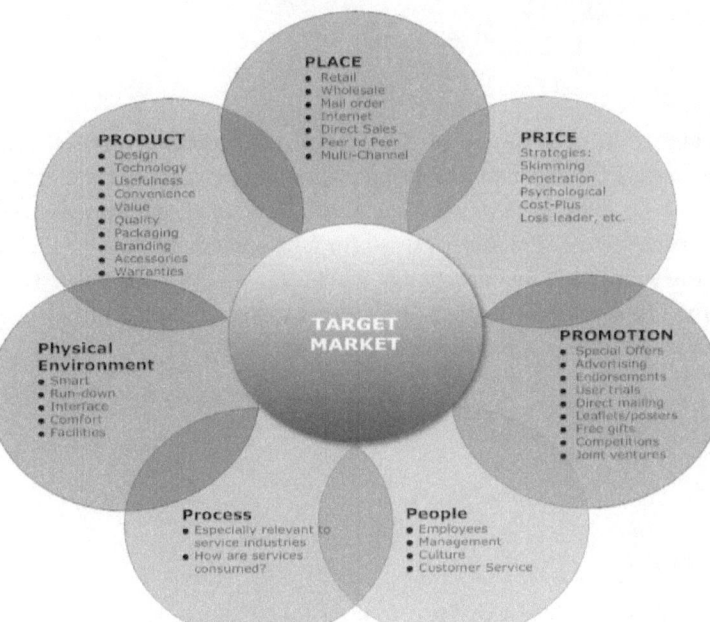

Price

Price is one of the most important factor which affects consumer's choice to purchase a particular product or brand. According to Law of Demand as the price of a commodity increases the demand for the same decreases. "Price is the amount of money charged for a product or service, or the sum of the values that customers exchange for the benefits of having or using the product or service". Price has been found to have a significant influence on purchase intension in many previous studies.

Product Features

Feature is an attribute of a product to meet the satisfaction level of consumers' needs and wants, through owning of the product, usage, and utilization of a product. Product features includes hardware and software. Hardware is the description for a device that can be touched physically. The hardware of a Smartphone is the body of the phone itself, size and weight. Colour and design are also considered as hardware as it is the physical appearance of the Smartphone. Software whereas is the general term for computer programs, procedure and documentation. The software of a Smartphone is the operating platform, storage memory, or apps that run the phone."

Compatibility

Compatibility is also an important issue which influences consumer's perception and purchase intension especially when the product of technical nature like Smart phone. Product compatibility is a unique outcome of symmetric perfect and firm should decide whether to make their product compatible before competing in prices . Qun, Howe, Thai, Wen and Kheng (2012) examined the impact of compatibility on purchase intension and found that compatibility significantly affects consumer's purchase intension of Smart phones and together with price and social influence explains 51.6% of total variance. also found that compatibility significantly affects purchase intension.

Convenience

Convenience in Smartphone may refer to the ability to use the Smartphone at anytime and anywhere, without having to port the Smartphone in a fixed workstation . Technology Acceptance model (Davis, 1989, Davis and Venkatesh, 1996) which was developed to explain who accept new technology recognized that the intension to accept a new technology is determined by perceived ease of use and perceived usefulness. Perceived ease of use in this model is a reflection of consumer's perception, to what extent the technology is convenient and easy to use. Previous studies have found convenience to be a significant deterministic factor which influences consumer's purchase intension.

Brand Name

Brand name is another important factor which affects consumer's purchase behaviour. According to the American Marketing Association, brand is defined as the "name, term, symbol, or design, or a combination of them intended to identify the goods and services of one seller or group of sellers and to differentiate them from those of competition". Brands are more than just names and symbols. It is also the element of relationship between company and customers "Brand names are valuable assets that help correspond quality and suggest precise knowledge structures which related to the brand". Importance of Brand name is shaping consumer's behaviour towards Smart phones has been recognized in the previous studies.

Relative Advantage

Relative advantage is the degree to which an innovation is perceived as better than the product it supersedes, or competing products . The nature of an innovation determines what specific type of relative advantages is important to the people, although the potential adopter's characteristics also affect which sub dimensions made up the relative advantages.

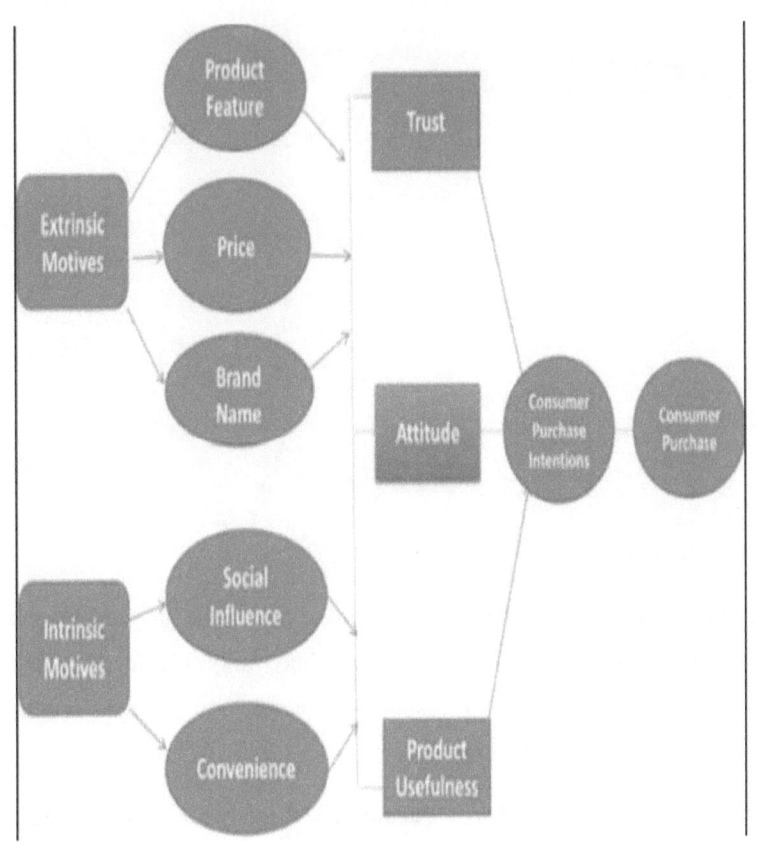

A Study of Consumer's Perception in Kolkata

Methodology Adopted

It is the process used to collect information and data for the purpose of making business decisions. The methodology may include publication research, surveys and other research techniques, and could include both present and historical information.

Sampling Technique/Design: Purposive sampling, also known as judgmental sampling, is one that is selected based on the knowledge of a population and the purpose of the study. The subjects are selected because of some characteristic. The idea behind purposive sampling is to concentrate on people with particular characteristics who will better be able to assist with the relevant research.

Sample Size: 94 respondents

Sampling Area: Kolkata, India

Sample Unit: All people of Kolkata city have been taken as sample unit who are well provided for smartphones.

Research Design: Descriptive research design

Research Instrument Used: Structured Questionnaire using 5-point Likert Scale

Data Collection Technique: In this project my primary data was questionnaires which were distributed to the consumers and data was collected according to that. Again, secondary sources were also used, which is nothing but the data collected and processed by some other person or organization. The sources of secondary data were official website of the company and various other publications, journals, magazines, newspapers and so on and reports present with the company. On the basis of information gathered, analyses on various points were made and recommendations and conclusions were drawn.

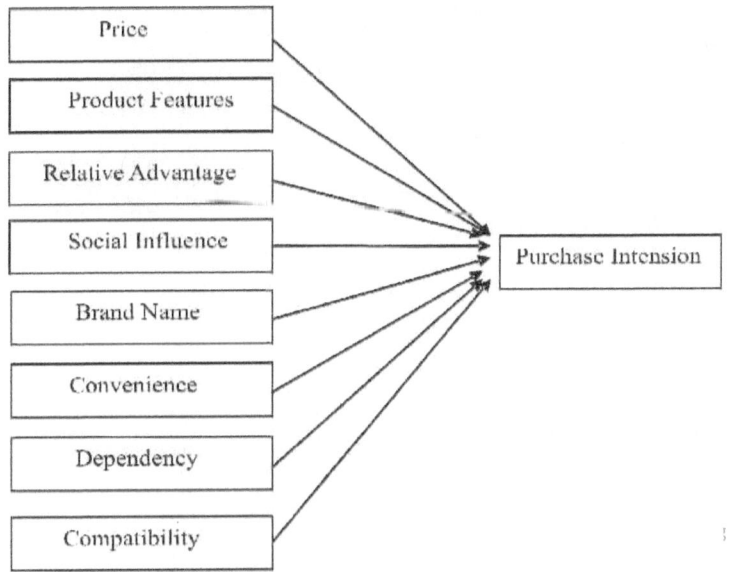

Research Gap

A research gap is defined as a topic or area for which missing or insufficient information limits the ability to reach a conclusion for a question.

In the context of this study, very few studies have been undertaken in Kolkata previously. The variables have not been studied together, thereby stressing my motivation to pursue this research.

Key Variables used in the Study

The Conceptual framework is based on the theoretical framework of this research study. It defines how the models in this research study are related to each other and gives a bird–eye view of the frame work which the conceptual model is based on.The various concepts used in this research like brand equity measures loyalty, awareness, association, perceived quality are all used in the questionnaire. While the Brand identity measures personality, physique, relationship, culture. The measurement of the various aspects of brand identity and brand equity asses how each contributes to brand building. The brand identity and brand equity models are link together by brand building model.

Brand

In today's marketplace teeming with thousands of products and services, all of which are being rapidly commoditized, a brand stands out from the clutter and attracts attention.

A brand name can create and stand for loyalty, trust, faith, premiumness or mass-market appeal, depending on how the brand is marketed, advertised and promoted. Brand differentiates a product from similar other products and enables it to charge a higher premium, in return for a clear identity and greater faith in its function. A brand is also likely to survive longer than just an undifferentiated product.It is akin to a living being: it has an identity and personality, name, culture, vision, emotion and intelligence. All these are conferred by the owner of the brand and needs to be continuously looked at to keep the brand relevant to the target it intends to sell to.

Kapferer (2004) defines brand as a name that influences buyers. He further notes that brand command people's attention because they have element of saliency, differentiability, intensity and trust. Successful brand conveys a consistent message and create an emotional bond with consumers. Hammond (2008, p.14) defines brand by " A brand is the total emotional experience a customer has with your company and its product or service" the author describes the brand to be an experience that's implanted in the mind of customers that have experienced an interaction with a company or that got in contact with the company's staff, product or service. In case the customer experience is different from what the company illustrate its brand. In that case the company is losing. However other parts of the brand such as "logos, advertising campaigns, mission statements, colors holds also great importance but they should always come after the customer, it is vital to put the customer and the company's relationship with him/her in the first position and think how to develop this relationship and then as a second stage a company can use all important tools like logo's, color's and advertising campaigns. The remarkable benefit that strong brand can bring to the company. Strong brand increases company's sales volume, it create great awareness and introduces the product or the service that the company is offering, in addition to that it builds a long-lasting customers relationship and loyalty, and in some cases it give the chance for the company to offer their product and services with a premium pricing. Companies with powerful brands also enjoy having devoted employees who believe in the brand and are loyal to it. On the other hand having a powerful brand becomes worth huge amount of money added to the company's balance sheet, for example Apple, IBM, Microsoft was worth around $80 million and above

Brand Equity

Brand equity is among the few strategic assets available to the companies that provide a long-lasting competitive advantage to the company. Brand equity constitutes the assets and the liabilities that is link to a particular brand, like name, or logo. It comprises of brand loyalty, brand awareness, brand association, brand assets, and perceived quality. Creating strong, favorable and unique brand association is a real challenge for markets but it is essential in building strong brand. Strong brands typically have firmly established strong, favorable and unique brand association with customer.It is a set of asset and legal responsibility connected to the brand's name and figure that add to (or take away from) the value presented by the product or service to a company and/or that company's customers.The main assets are grouped in the followings:

• Brand name awareness

• Brand Loyalty

• Perceived quality

• Brand associations

Aaker (1996, p 8) demonstrate in the below figure how brand equity generates value, through investment these assets can be created and improved. Each one of the five brand equity assets produces value in a diverse way; seventeen of these values are listed in the figure below. It is essential to be aware of the ways in which well-built brands forms value, in order to take important decisions about brand building activities.Brand equity forms value for both the customer and the firm; however the brand equity should be expressed and linked by the name and symbol of the brand. Aaker (1991,p.15) define a brand as a distinguishing name or symbol intended to identify the goods or service of either one seller or group of seller and to differentiate those goods or service of product and protect both the customer and the producer and product from competitors. A strong brand can command price premiums; on the other hand strong brand cannot command an excessive price premium. Brand equity occurs when consumer response to marketing activity differs when consumer know the brand from what they do not. Brand equity can be positive or negative, positive brand equity is often when a company exceeds their costumer expectation. It formed by past efficient and successful advertising. Negative brand equity is created by usually bad management. Positive brand strategy usually is a strong barrier to entry for potential rival and competitor. The more a brand is known, the more its advertisement are noticed and remembered. Brand equity can provide the company with a high potential of future growth.

The consumers are willing to pay high price companies with brand. Brand equity such as brand awareness, image, trust, and reputation are painstakingly built up over many years. Branding requires a long term corporate involvement, a high level of resource and skills. Brand equity incorporates several advantages like, it ability to attract new customers, resist competitive activity, lower advertising /sale ratios, brand loyalty, trade leverage and premium pricing. Brand allows consumers to lower search costs for product both internally (regarding how much they have to think) and externally (in terms of how much they have to search around). It also help consumers make assumption of the product base brand quality and characteristics, thereby forming a reasonable expectation about what they do not know about the brand. It conveys a certain level of quality so that satisfied buyer can chose the product again. Brand equities is said to be the most intangible assets that a company posses that is not reflected in the balance sheet, given that it provide competitive advantage for the company. Having a brand guarantees future earnings. Despite the fact that the corporate balance sheet display only assets of the company without mentions of the brand equity does not change the fact that it is precious assets to the company. While a competitor can copy a company's product but its brand is unique.

Brand Loyalty

Brand loyalty is often ascribe to a behavioral sense through the number of repeat purchases, it entails consumer sticking with the brand and reject the overture of competitors. Strong brand equity holds consumer loyalty because consumer values the brand on the basis of what it is and what it represents. Brand loyalty is attributed to brand image and brand equity. It is also worth noting that brand commitment is the substantial expression of brand preference and brand loyalty. Brand loyalty ascertains, that extent the customer is attached to a brand and speculate how likely the customer will switch to another brand, when the brand changes either in product price or features. Brand loyalty of current customers represents a strategic assets, and when properly managed would provide the company with several values. Brand loyalty it associated more closely to the use experience, in the sense that it does not exist without prior purchase and use experience.

There are five different level of brand loyalty; each level poses different challenges and different assets to companies. The bottom loyalty level is the nonlocal buyers who have no preference for any brand. These groups of buyers place no value on the brand name and consider any brand as adequate. Brand does not play any role in the buyer purchase decision. The second level includes buyers who are satisfied with pleased with the product. These groups of buyer are vulnerable to a competitor who offers them a visible benefit to switch; hence they are classified as habitual buyers. The third levels are class of buyers who are please with the brand, these buyer might switch brand if competitors overcome the switching cost by inducing the buyer to switch to their brand. The fourth levels of brand loyalty are the category of loyal customer that really likes the brand. Their preference for the brand is on the symbol, previous experiences or the logo of the company. The top levels are category of buyers who are committed to the brand, because of the brand functionality or because the brand is the expression of whom they are.

Smartphone Brand Loyalty

However Phones Review (2012) conducted voting hype for people to share their idea and what would they rather purchase an iPhone 5 or Samsung

Galaxy S3, and by going throw the comments that many people posted, we can note that Samsung has succeeded in building a strong brand loyalty and brand awareness, as many people were loyal to that brand and are willing to purchase it upgraded devices.

According to Cush (2010) Samsung had many happy customers in the U.S., Samsung closed 2009 as the No. 1 handset market share holder, further to the Strategic Analytics' also no. 1 in cell phone marker in 2010. Brand Keys Customer Loyalty Engagement Index, which measures clients favorite mobile devices design and performance according to identify brands that meet up or exceed consumers outlook. The Brand Key national survey for Smartphone and standard cell phone, showed that Apple ranked no.1 in the Smartphone categories, in second comes Samsung and then followed by Blackberry. As shown in Figure , Retention Rate, Rosati (2011) demonstrate that in the survey conducted by UBS in the August 2011 to 515 consumers to find out how many of them will continue in buying smart phones made by the same company which they bought their current device from and to determine how many of them would want to change to a different manufacture. As shown in Figure 3, 89% retention rate for Apple consumer, they said that they will buy their upgraded device from the same company (Apple) which puts Apple in No 1 position for customer loyalty and satisfaction. In the second position ranked HTC but still far away from the percentage the Apple had, 39% retention rate, RIM comes next with a percentage of 33%, Samsung had ranked in the fourth position with 28% retention rate. Behind comes Motorola with 25% and Nokia 24%. In addition to the above the study showed that 31% of the Android users were more likely to switch to buy an I Phone for their next device.

Brand Awareness

Awareness is the power and force of the brand's presence in the consumer's mind, consumer, and this awareness is measured by how consumer keeps in mind the brand and how they recall it.

Recognition of a brand is based on the knowledge added from earlier exposure, it is not essential to remember where the brand was seen or why it is diverse from other brands neither it needs to be remembered the class of the brand, remembering the earlier exposure is the most important thing. Economist explains that when costumers recall being exposed to a brand this will be transmitted in their brain to be a "signal" of a good brand because they believe that companies don't pay out large amount of money on bad product.

The crucial part of brand awareness is the brand name dominance. This is why companies should start protecting at the beginning of its existence. When choosing a brand name firms should beware of descriptive names, such as Windows because this makes it harder for customers to differentiate from the generic product. Due to the fact that customers are being exposed to more and more brands and messages, creating brand awareness is a big challenge

The two way to create the awareness:

- First is by creating "healthy awareness levels" this wide trade base is typically an asset, but it is expensive and for small unit sales and for life measured in years and not in decades this can be impossible. This is why huge companies have an advantage in building the awareness. Multiple businesses sustain and back up the brand name.
- Second in future companies will experts in working outer the usual media channels, by using occasion promotions, sponsorships, advertising and sampling, these companies will be the most successful companies in creating the brand awareness. However, companies should always put in mind while managing the brand that it is necessary that brands are managed not for "general awareness" but for "strategic awareness" it is important that customers recall this brand for good reasons and not for bad ones.

Smartphone Brand Awareness

Further to Pomoni (2010) article, Apple is one of the brands that have strong brand awareness; nearly 90% of consumers are aware of it.Which gives the opportunity for Apple to present consumers with different products with an effective position, which increase's Apple market shares because many consumers will choose Apple products over a similar product of different company because of the company high awareness. According to Fee logo Services (2006) one of Apple awareness tools is the logo the smart and simple design of the logo, made this logo the most recognized brand symbol all over the world, it has demonstrated that is effective and easy to remember.

Famous logos (2012) clarifies the phase that Apple gone through in creating its current logo, as shown in the below figure (figure 4) the design that Apple logos have now is after a long and deep thinking, a story is behind this logo evolution. Sir Isaac Newton, who created the law of gravity, was sitting under a tree and an Apple fell down which inspired him in finding this theory of law gravity. The logo of Apple is clearly connected to this Apple. At the beginning the logo was designed in having a rainbow color which represented the new creation of colors in the IT world.

Some people believe that logo is also connected to Alan M. Turing, who is considered to be the father of computer science, who gave end to his life by eating a poisoned apple. Now the logo of Apple is colored in polished silver adding a modern touch to the logo. This logo is only one of its kinds and holds an fascinating background and story. It is easy to understand and signifies the best quality of products

Famous logos (2012) Samsung logo color, blue mean honesty, trust and reliability and dedication in service.

Two character distinguish the Samsung logo from other companies logos, first is the figure is turn around ten degrees from the x-axis, which presents a sense of dynamic tension. The second character is the brand's name "Samsung" which flashes out from the figure, this clearly dissect it and give the impression as if the out shape is a spotlight for the clearing and defining the brand name.

According to the report provided by Mike Luttrell Samsung aggressive brand recognition campaigns has earned them huge brand awareness among consumers. The research report conducted shows that 54% of prospective Smartphone buyer discloses that they would go for iPhone. Recently there has been 5% increase in consumers interested in Samsung Smartphone.

Samsung embarked on sponsorship in order to create brand awareness. It sponsored Athens 2004 Olympic game which contributed positively to the brand awareness, giving it a boost from 57% to 62%. Also the sponsorship help it to build a good reputation and brand image. The awareness rise leaded to the sales surge.

Brand Identity

Brand identity stems from an organization, i.e., an organization is responsible for creating a distinguished product with unique characteristics. It is how an organization seeks to identify itself. It represents how an organization wants to be perceived in the market.

Kapferer (2004, p.107) postulated the brand Identity Prism. It is a hexagonal prism which comprises of six facets of brand identity namely Physique, personality, relationship, reflection, and self-image. The brand identity prism illustrates that these facets are all interrelated and form a well-structured entity. The significance of the brand prism is to help in understanding the essence of brand and retailer identities.

Brand is a physique. A physique deal with the physical aspect of the brand .It is a combination of the prominent objective features which comes to mind when the brands is mentioned. It represents the backbone and its tangible added value. The physical aspect of the brand is one of the things that define a brand.

Brand is a culture. In the sense that it has its own separate culture which is derive from products. A product embodies not only culture but also is a means of communication. Brand conveys culture and is driven by culture in the sense that they convey the culture of the society they originate from. Country of origin is the reservoirs for brands. One vital role of culture is that it link brand to the firm and play essential role in differentiating brand.

Brand is a relationship: This facet of brand ,has implication for the way the brand act ,delivers service ,or relate to its customer .Brand are sometimes the crux of transactions and exchange between people. Brand relate its name to a theme .E.g. Like Nike beer a Greek name that relate it to specific cultural values, to the Olympic games. Kapferer (2004, p. 107).

Brand has a personality. People usually ascribe brand with certain personality. The way in which it speaks of its product or service shows what kind of person it would be if it were human. It also pointed out that brand personality is described and measured by those human personality traits that are relevant for brands. Brand is a customer reflection. Brand reflects the individual who use it. A brand always tends to reflect or build an image of the buyer or the user which it seems to be addressing. It does reflect the customer as he or she wishes to be seen as a result of using a brand. Consumers use brand to build their own identity. Customers want to be seen in certain way as a result of purchasing a given product.

Everything a brand does is communication. "It is impossible not to communicate." The way the packs are designed, the words used, the way the phones are answered (or not), and the products the name is put to, the shops in which these are sold: all these can say powerful things about a brand. Brand speaks to our self-image. This face of brand identity deals with our relationship the inner self-image that is built through our attitude towards certain brand.

Brand Personalities

Aaker (1996, p. 141) defines the brand personalities to be "the set of human characteristics associated with a given brand". Thus it includes such characteristic as gender, age, and socioeconomic class, as well as such classic human personality traits as warmth, concern, and sentimentality. Brand personalities are similar to human's personalities; they are different and permanent or long lasting.

As mentioned before by Aaker (1996, p. 142) brand personalities can be described by the following:

• Described by demographics such as: age, gender, social class and race.

• Described by life style such as: activities, interests and opinions.

• Described by human personality traits such as: extroversion, agreeableness and dependability.

Smartphone Brand Personalities

CBC News (2008) noted that the online magazine brand channe.com, which is specialized in Studying trends and brand awareness, conducted an online survey for around 2000 professionals and students from 107 country around the world, where they asked them which brand inspires them the most, 22% of the answers replied that Apple inspired them the most, this is why Apple ranked top marks. When people where asked about the importance of brands, by asking them "what brand would you most like to sit next to at a dinner party?" and "if you were to describe yourself as being a brand, what brand would you be?" Apple brand marked in the second position for that answer.

Marketing Minds (2012) noted "the Apple branding strategy focuses on the emotions and feelings, the brand personality is all about the lifestyle; imagination; liberty regained; innovation; passion; hopes, dreams and aspirations and power-to-the-people through technology. The Apple brand personality is also about simplicity and the removal of complexity from people's lives; peopledriven product design; and about being a really humanistic company with a heartfelt connection with its customers." Apple has created a strong brand that represents stylish technology and high price products. It core values embodies innovation, stylishness in product design and ease of use which attribute immensely to the image of the brand. Apple is a master creating strong brand that achieved incredible brand loyalty. Even though Apple is a technology company Apple brand emerges into a lifestyle

Marketing Mix / Marketing Strategy

Marketing mix refers to the set of actions, or tactics, that a company uses to promote its brand or product in the market. The 4Ps make up a typical marketing mix - Price, Product, Promotion and Place. However, nowadays, the marketing mix increasingly includes several other Ps like Packaging, Positioning, People and even Politics as vital mix element.

All the elements of the marketing mix influence each other. They make up the business plan for a company and handled right, can give it great success. But handled wrong and the business could take years to recover. The marketing mix needs a lot of understanding, market research and consultation with several people, from users to trade to manufacturing and several others.

DATA ANALYSIS

AGE (94 responses)

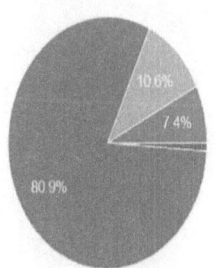

- Below 18
- 18-25
- 26-35
- 35-50

10.6%
7.4%
80.9%

OCCUPATION (94 responses)

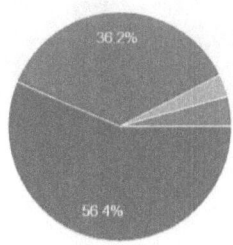

- Student
- Proffesional
- Buisnessman
- Other

36.2%
56.4%

Scale for the Measurement

- **1-Highly Disagree**
- **2-Disagree**
- **3-Neutral**
- **4-Agree**
- **5-Highly Agree**

Reason for switch:-

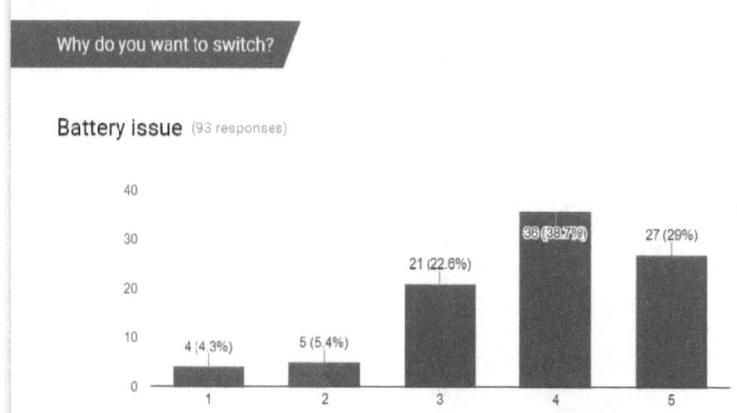

Broken Screen (92 responses)

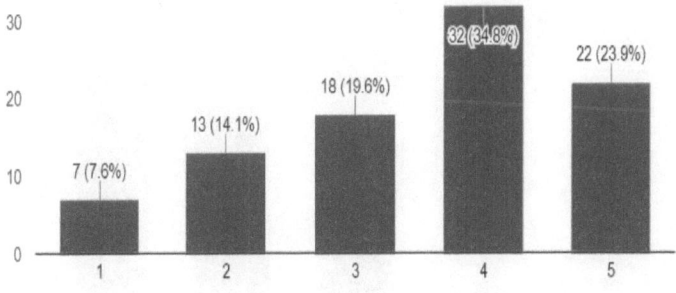

Dents (90 responses)

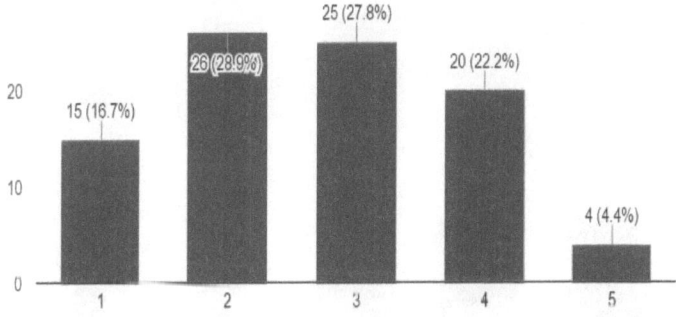

Better Features (93 responses)

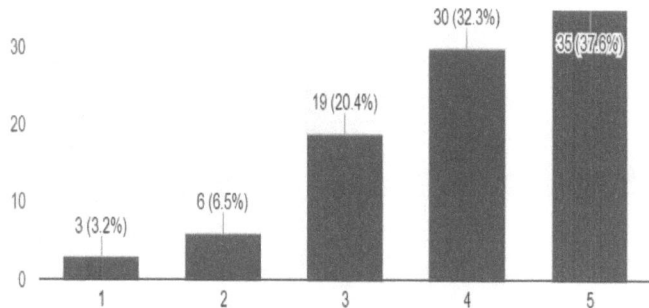

Bored with current phone (93 responses)

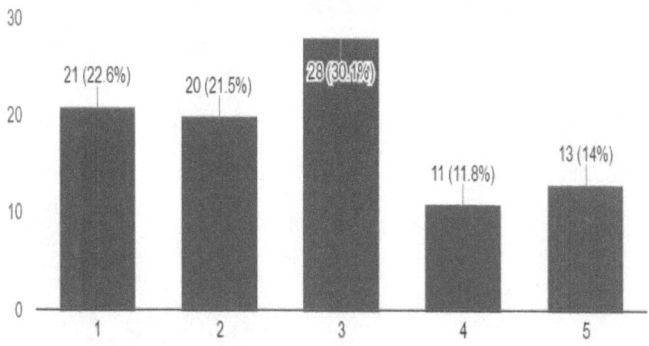

Prioritize in purchase decision

Elements you will prioritize when making a purchase decision

Price (93 responses)

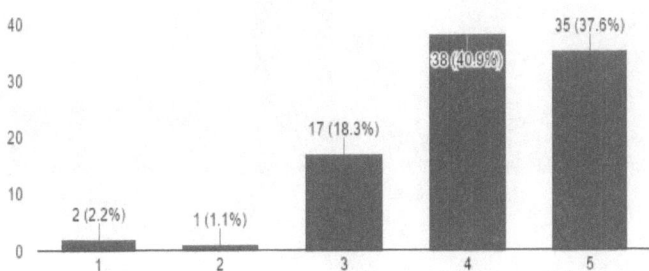

Trends (92 responses)

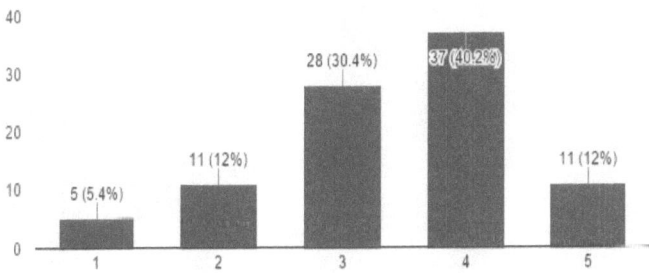

Aesthetics (90 responses)

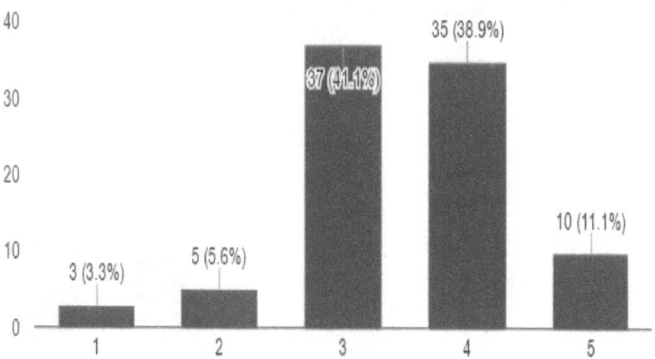

RAM (92 responses)

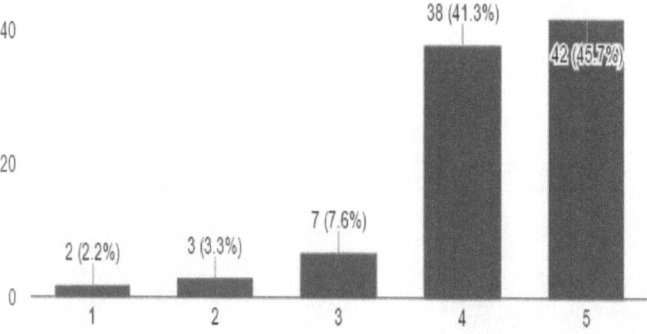

Consumer Dynamics

I have a high level of involvement (in terms of research activities undertaken)
(94 responses)

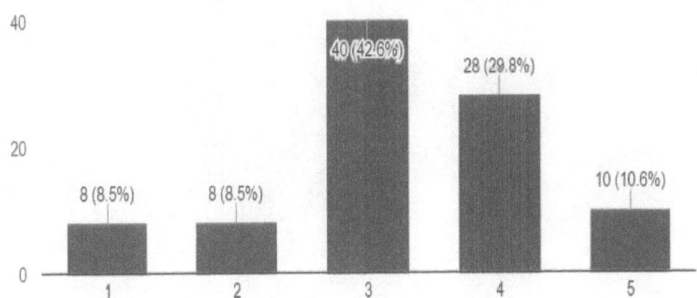

Which brand of smartphone do you use? (94 responses)

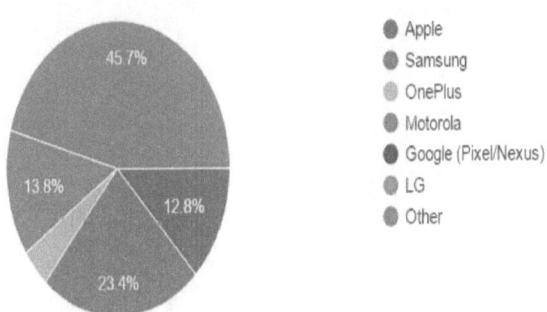

If you switch, which brand will you switch to? (94 responses)

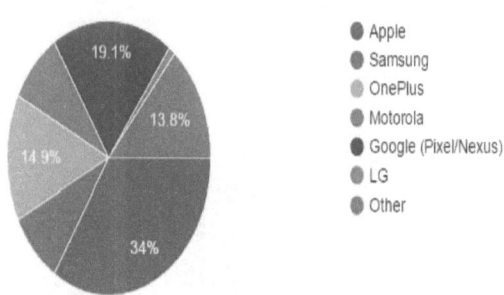

How likely are you to switch your perfectly functional smartphone for a newer, better one?
(94 responses)

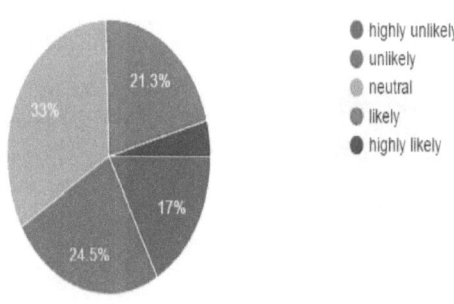

Limitations

Like any other research study the present study is also not free from limitations. For the present study a sample of 94 respondents was taken, though the given sample is sufficiently large for the study still a bigger sample could lead better results. Due to time, budget and other limitations a non probability convenience sampling was used to collect the data therefore the data may or may not be fully representative of the given population which may affects the generalizability of the results of the study. In short

- → Time was the main issue. One month time was not enough to study the whole region.
- → In some cases the respondents were also not in a mood to answer the questions, which affected the accuracy of the questionnaire.
- → As the sample size was small, exact information was difficult to achieve.

Conclusion

Presently when technology is changing rapidly , mobile phone is not just a device which is used for calling and messaging rather it is now called as Smart phone because of its increasing use in day today life. It has become very difficult to imagine life without Smartphone. The present study has explored that there are eight major factors and they are price , product features, relative advantage, convenience, compatibility, social influence, dependency and brand name which influence consumer intension to purchase Smartphone. However only social influence, compatibility and dependency were found to be significantly influencing purchase intension of the Smart Phone consumers. Though rest of the factors are also important but these three factors are major reason why a consumer wants to purchase Smartphone .

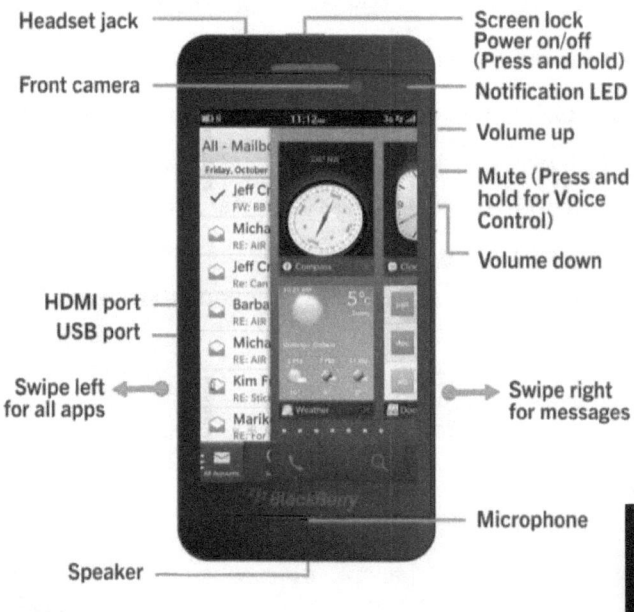

www.ingramcontent.com/pod-product-compliance
Lightning Source LLC
Chambersburg PA
CBHW030539220526
45463CB00007B/2905